Bibliographic information published by the German National Library:

The German National Library lists this publication in the National Bibliography; detailed bibliographic data are available on the Internet at http://dnb.dnb.de .

Imprint:

Print and binding: Books on Demand GmbH, Norderstedt Germany
ISBN: 978-3-668-04792-1

This book at GRIN:

http://www.grin.com/en/e-book/305585/multiple-choice-questions-for-biological-sciences

Vijay Kothari

Multiple Choice Questions for Biological Sciences

GRIN Publishing

Multiple Choice Questions for Biological Sciences: An aid for the aspirants of various competitive examinations

Vijay Kothari Ph.D.

Institute of Science
Nirma University
Ahmedabad
India.

This is a compilation of more than 100 multiple choice questions pertaining to different areas of biological sciences. This compilation is intended to be helpful to those who are preparing for appearing in any of the competitive examinations at various levels. Questions mainly are from the fields of Microbiology, Biochemistry, Biotechnology, Immunology, Biomedical Engineering, etc. All correct answers are put in bold face for immediate reference of the reader. Teachers may also find some questions from this compilation suitable for inclusion in various test papers.

(Correct answers are highlighted in bold face)

1. Introns are absent from genome of :

A. Eukaryotes

B. Prokaryotes

C. Both

2. Which of following is a non-polar solvent:

A. Methanol

B. Chloroform

C. Ethanol

3. Which of the following branches deals with biochemical pathways of plants?

A. Phytochemistry

B. Pharmacognosy

C. Plant Biochemistry

4. The first human disease proved to have a viral cause was_________

A. Smallpox

B. Rabies

C. Hepatitis

D. Yellow fever

5. Dark ground microscopy is used for detection of _________

A. Spirochetes

B. Fungi

C. Chlamydia

D. Virus

6. Bacteria with tuft of flagella at one end are called_________

A. Monotrichate

B. Peritrichate

C. Bipolar

D. Lophotrichate

7. Which of the following statements about endotoxins is false?

A. They are heat stable.

B. Protein-Polysaccharide-lipid complex in nature

C. Action is often enzymatic

D. Is produced by gram-negative bacteria

8. Hapten _______

A. Produces humoral response

B. Binds to carrier to produce its effects

C. Is high molecular weight protein

D. Is the same as epitope

9. Which immunoglobulin can cross the placenta?

A. Ig A

B. Ig G

C. Ig D

D. Ig E

10. Widal test is a type of _______

A. Precipitation reaction

B. Agglutination reaction

C. Complement fixation test

D. Immunofluorescence

11. Diphtheria toxin acts by__________

A. inhibiting acetylcholine release

B. Inhibiting glucose transport

C. Increasing levels of cyclic AMP

D. Inhibiting protein synthesis

12. The virulence of gonococci is due to________

A. Pili

B. Cell membrane

C. Its intracellular location

D. Cytolytic enzymes

13. What differentiates Nocardia from Actinomycetes?

A. Gram stain

B. Modified acid-fast stain

C. Growth on Sabouraud's medium

D. Animal pathogenicity

14. Segmented RNA is observed in

A. Rabies virus

B. Influenza virus

C. Coxsackie B virus

D. HIV

15. Which of the following conditions is not of endogenous origin?

A. Phycomycosis

B. Aspergillosis

C. Candidiasis

D. Both A and B

16. Which of the following is associated with dandruff?

A. Candida albicans

B. Malassezia furfur

C. Propionibacterium acnes

D. Clostridium botulinum

17. The only deep mycosis common in India is __________

A. Cryptococcosis

B. Coccidiodomycosis

C. Histoplasmosis

D. Blastomycosis

18. Which of the following is not true.

A. Cells in biofilm exhibit greater antibiotic resistance

B. Cells in a biofilm exhibit a slower pace of metabolism and replication

C. Bacterial population inside a biofilm reaches higher cell density than the planktonic forms

D. Biofilms are not involved in human infections.

19. Gene expression profile can be investigated through_________.

A. Genomics

B. Proteomics

C. Metabolomics

D. Transcriptomics

20. *Cag* pathogenicity island is present in________.

A. Vibrio cholerae

B. Helicobacter pylori

C. Streptococcus mutans

D. Staphylococcus aureus

21. Xanthan gum is obtained from which of the following microbes?

A. Xanthomonas citri

B. Citrobacter spp.

C. Erwinia carotovora

D. Xanthomonas campestris

22. Which of the following mode of microbial communication is likely to be faster, and more effective for less denser populations?

A. Quorum sensing making use of chemical signals

B. Communication using physical signals such as sound

C. Both will happen at almost same speed

D. None of the above

23. Vibrations can be detected by microbes using__________.

   a. Mechanosensory channels

   b. Cell wall

   c. Cell membrane

   d. Intracellular fluid

24. Which of the following is a gram-positive bacterium?

a. E. coli

b. S. mutans

c. S. typhi

d. V. cholerae

25. Which of the following is a gram-negative bacterium?

a. S. mutans

b. B. subtilis

c. S. aureus

d. V. cholerae

26. Role of a buffer is to maintain _________.

a. pH

b. Temperature

c. Pressure

d. pH and temperature

27. Beer-Lambert law is relevant to which of the following?

a. Spectrometry

b. Spectroscopy

c. Chromatography

d. Ultrasonication

28. Which of the following is a respiratory pathogen?

a. S. aureus

b. V. cholerae

c. M. tuberculosis

d. All of the above

29. Which of the following is an eukaryotic organism?

a. Candida albicans

b. Saccharomyces cerevisiae

c. *Plasmodium falciparum*

d. All of the above

30. Malaria is caused by ________.

a. Mycobacterium spp

b. Plasmodium spp

c. *Acinetobacter spp.*

d. Basidiomycetes

31. A 12% solution of Nacl will contain ____ g of Nacl in 100 mL of solvent.

a. 12

b. 120

c. 1.2

d. 0.12

32. 1 μg/mL is equivalent to _______ mg/L.

a. 0.01

b. 0.1

c. 10

d. 1

33. For which of the following wavelength ranges a quartz cuvettee has to be used?

a. 400-700 nm

b. 450-800 nm

c. 200-350 nm

d. None of the above

34. HEPA stands for__________.

a. Highly effective particle aid filter

b. High efficiency particulate air filter

c. Heavy electron parity acceleration filter

d. None of the above

35. Which of the following is not having a cell wall ?

a. Animal cell

b. Plant cell

c. Bacterial cell

d. Algal cell

36. Which of the following is most suitable for growth of fungi?

a. Nutrient broth

b. Urea broth

c. Potato dextrose broth

d. All of the above

37. Which of the following can be used as a differential media for Enterobacteriaceae?

a. Sellenite broth

b. Muller-Hinton broth

c. MacConkey's broth

d. Urea broth

38. In a plug flow reactor, the highest concentration of substrates is exposed to the cells that-

A. are near the effluent or exit of the reactor

B. are in the mid way along the reactor

C. are located near entrance of feed

D. are near the wall of the reactor

39. A strain of *Escherichia coli* has a maximum specific growth rate of 0.8 h^{-1} on a glucose based medium. If this organism is being grown in a chemostat with a dilution rate of 1.2 h^{-1}, then at steady state the concentration of *E. coli* in the same medium will

A. increase

B. be zero

C. change randomly

D. decrease

40. Yeast is typically produced in fed batch reactors because

A. yeast cells produce ethanol at low glucose concentrations

B. yeast cells ferment at low glucose concentrations

C. **yeast cells respire at low glucose concentrations**

D. all of the above

41. In a fluidized bed reactor mixing is achieved by

A. gravity flow of fluid

B. upward movement of the incoming feed

C. diffusion

D. **both (a) and (b)**

42. In fed batch bioreactor modelling, the rate of change in the bioreactor volume is assumed to be equal to

A. **flow rate**

B. initial volume

C. volume of solids in the reactor

D. none of these

43. Production of organic acids in batch culture is generally growth associated phase followed by non-growth associated production. The probable reason for this is that

A. **organic acids uncouple catabolism from anabolism**

B. biomass yields increase as the fermentation proceeds

C. organic acids are secondary metabolites

D. high concentrations of organic acid promote the growth of the microbial population

44. When the product formation is approximately equal to the rate of cell growth, the pattern of product formation is termed as

A. uncoupled

B. growth associated

C. non-growth associated

D. metabolically uncoupled

45. Which of the following would be correct concerning the concentration of biomass (X). growth limiting substrate (S) and product (P) in a chemostat when the cells have been washed out?

A. X, S and P will be zero

B. X, S and P cannot be predicted for the above condition

C. **X and P will be zero while S is equal to the substrate concentration in feed**

D. None of the above

46. *Saccharomyces cerevisiae* when grown in high glucose concentration medium, the cells activity depends on which of the following?

A. High concentration of dissolved oxygen

B. Low concentration of dissolved oxygen

C. **No effect of dissolved oxygen**

D. High concentration of dissolved oxygen and sodium salts

47. In the batch-fed mode, substrate will be added

A. only at the beginning of the fermentation process

B. **periodically throughout the fermentation process**

C. only at the end of the fermentation process

D. substrate is never added

48. Secondary metabolites are also known as:

A. Natural Products

B. Non-essential metabolites

C. Green metabolites

D. None of the above

49. Introns are present in the genome of :

A. **Eukaryotes**

B. Prokaryotes

C. Both

D. None of the above

50. Which of following is a non-polar solvent:

A. Methanol

B. Hexane

C. Ethanol

D. Water

51. Which of the following branches deals with medicinal properties of plants?

A. Phytochemistry

B. Pharmacognosy

C. Plant Biochemistry

D. Plant Biotechnology

52. Which of the following plant metabolites are most susceptible to oxidation?

A. Phenolics

B. Alkaloids

C. Terpenoids

D. None of the above

53. Gram-staining is a __________ type of staining.

A. differential

B. simple

C. monochrome

D. None of the above

54. Which of the following is a gram-positive bacterium?

A. Escherichia coli

B. Clostridium botulinum

C. Salmonella typhi

D. Vibrio cholerae

55. Which of the following is a gram-negative bacterium?

a. S. mutans

b. B. subtilis

c. S. aureus

d. ***V. cholerae***

56. Role of a buffer is to maintain ________.

  A. pH

  B. Temperature

  C. Pressure

  D. pH and temperature

57. Beer-Lambert law is relevant to which of the following?

  A. Spectrometry

  B. Spectroscopy

  C. Chromatography

  D. Ultrasonication

58. Which of the following is a respiratory pathogen?

  A. S. aureus

  B. V. cholerae

  C. M. tuberculosis

  D. All of the above

59. Which of the following is an eukaryotic organism?

  A. *Candida albicans*

  B. *Saccharomyces cerevisiae*

  C. *Plasmodium falciparum*

  D. All of the above

60. Malaria is caused by ________.

  A. *Mycobacterium spp*

  B. Plasmodium spp

  C. *Acinetobacter spp.*

  D. Basidiomycetes

61. A 12% solution of Nacl will contain ____ g of Nacl in 100 mL of solvent.

A. 12

B. 120

C. 1.2

D. 0.12

62. 1 µg/mL is equivalent to _______ mg/L.

A. 0.01

B. 0.1

C. 10

D. 1

63. For which of the following wavelength ranges a quartz cuvette has to be used?

A. 400-700 nm

B. 450-800 nm

C. 200-350 nm

D. None of the above

64. HEPA stands for__________.

A. Highly effective particle aid filter

B. High efficiency particulate air filter

C. Heavy electron parity acceleration filter

D. None of the above

65. Which of the following is not having a cell wall ?

A. Animal cell

B. Plant cell

C. Bacterial cell

D. Algal cell

66. Which of the following is most suitable for growth of fungi?

A. Nutrient broth

B. Urea broth

C. Potato dextrose broth

D. All of the above

67. Which of the following can be used as a differential media for Enterobacteriaceae?

A. Sellenite broth

B. Muller-Hinton broth

C. MacConkey's broth

D. Urea broth

68. Dark ground microscopy is used for detection of _________

A. Spirochetes

B. Fungi

C. Chlamydia

D. Virus

69. The gram-positivity of bacteria depends upon_________

A. Pore size of cell wall

B. Presence of mycolic acid

C. Presence of dipicolinic acid

D. Presence of teichoic acid

70. Interleukin-1 is produced by______

A. T-helper cells

B. B-helper cells

C. Monocytes

D. None of the above

71. Staphyloccocal toxic syndrome is due to___________

A. Enterotoxin A

B. Enterotoxin E

C. Enterotoxin D

D. Enterotoxin F

72. Which of the following is not caused by a virus?

a. Tetanus

b. Rabies

c. Hepatitis

d. Yellow fever

73. Bacteria with tuft of flagella at one end are called_________

a. Monotrichate

b. Peritrichate

c. Bipolar

d. Lophotrichate

74. Which of the following statements about endotoxins is false?

a. They are heat stable.

b. Protein-Polysaccharide-lipid complex in nature

c. Action is often enzymatic

d. Is produced by gram-negative bacteria

75. Which immunoglobulin can cross the placenta?

a. Ig A

b. Ig G

c. Ig D

d. Ig E

76. Which of the following is the transport medium for cholera?

A. Tellurite medium

B. Chacko-Nair medium

C. Venkatraman-Ramakrishnan medium

D. McLeod's medium

77. Which of the following is not an antibacterial drug?

A. Amphotericin B

B. Nystatin

C. Ampicillin

D. Both a & b

78. Which of the following can be involved in horizontal gene transfer (HGT)?

A. Conjugation

B. Transformation

C. Transduction

D. All of the above

79. Which of the following is used to treat HIV infection?

A. Tamiflu

B. Zidovudine

C. Sterptomycin

D. Ketoconazole

80. Which of the following conditions is of endogenous origin?

A. Phycomycosis

B. Aspergillosis

C. Candidiasis

D. All of the above

81. Pyocyanin and pyoverdin are the metabolites produced by which of the following organism?

A. *Vibrio cholerae*

B. *Pseudomonas aeruginosa*

C. *Streptococcus mutans*

D. *Staphylococcus aureus*

82. Which of the following is most commonly associated with dental caries?

a. *Porphyromonas gingivalis*

b. *Streptococcus mutans*

c. *Alternaria alternata*

d. *Malassezia furfur*

83. Which of the following is a non-invasive mucosal pathogen of the gastrointestinal tract?

a. *Vibrio choleare*

b. *Candida albicans*

c. *Helicobacter pylori*

d. All of them

84. Which of the following bacterium is most widely used as a model for quorum-sensing studies?

a. *Pseudomonas oleovorans*

b. *Chromobacterium violaceum*

c. *Xanthomonas campestris*

d. *Bacillus subtilis*

85. Physical vibrations are sensed by microbes using ________.

a. Mechanosensory channels

b. Flagella

c. Pili

d. Magnetosomes

86. Which of the following are cell-wall less organisms?

a. Mycoplasma
b. Thermoplasma
c. Mycobacteria
d. Both a and b

87. Which of the following is most suitable for cleaning of cuvettes?

a. Acetone
b. 0.5% detergent solution
c. Chloroform
d. Benzene

88. Optical Density (OD) = 2 – log __________.

a. Antilog OD
b. Transmission (T)
c. Antilog T
d. None of the above

89. NMR (nuclear magnetic resonance) is most closely associated with which of the following?

a. MRI (magnetic resonance imaging)
b. CT scan
c. Raman spectroscopy
d. All of the above

90. IR spectroscopy is complementary to______.

a. Mass spectrometry
b. NMR
c. X-ray crystallography
d. Raman spectroscopy

91. The ability of a microscope to distinguish between two nearby objects in the field is called_______.

   a. Magnification
   b. Compounding power
   c. Resolution
   d. Refractive index

92. Which of the following grows at the slowest pace?

   a. Bacteria
   b. Yeast
   c. Mold
   d. Lichen

93. Which of the following grows in filamentous form?

   a. Yeast
   b. Actinomycetes
   c. Mold
   d. Both b and c

94. Which of the following possess no cell wall?

   a. Bacteria
   b. Plant cells
   c. Animal cells
   d. All of the above

95. The triple screening test for Down syndrome performed between 15-20 weeks of pregnancy combines which of the following with blood tests of estriol, hCG and alphafoetoprotein?

(A) **Ultrasound** (B) Infrared

(C) MRI (D) NMR

96. When the cells from the amniotic fluid are cultured and tested for chromosomal abnormalities, this is known as:

A. Amniocentesis

B. Karyotyping

C. Amniotyping

D. Karyocentesis

97. Ultrasound scan at 11-14 weeks of pregnancy to check the *nuchal translucency* measures the thickness of the __________ at the back of the neck of the foetus.

A. Tissue

B. Bone

C. Muscle

D. Skin

98. Which of the following is not a type of foetal measurements by ultrasound?

A. Crown-rump length

B. Biparietal diameter

C. Femur length

D. **Thickness of amniotic fluid**

99. Which of the following imaging technique is used to measure the flow of blood in the blood vessels ?

A. **Doppler ultrasound**

B. 3-D ultrasound

C. Dynamic 3-D ultrasound

D. None of the above

100. *Cleft lip*- a type of foetal abnormality can be clearly defined using which of the following?

A. Doppler ultrasound

B. **3-D ultrasound**

C. Dynamic 3-D ultrasound

D. Foetal echocardiography

101. Which of the following ultrasound scan is particularly helpful in assessing suspected congenital heart defects?

A. Doppler ultrasound

B. 3-D ultrasound

C. Dynamic 3-D ultrasound

D. Foetal echocardiography

102. It is easier to see the baby more clearly in a thin mother with plenty of amniotic fluid, because-

A. **Increased abdominal fat can make it difficult for the sound waves to penetrate**

B. The statement of the question is not correct

C. Thickness of abdomen has nothing to do with an ultrasound

D. None of the above

103. Foetoscope is used to-

A. measure blood pressure

B. **listen to the foetal heartbeat**

C. measure width of the heart valve

D. None of the above

104. Pressure in arteries when heart contracts is:

A. **110**

B. 80

C. 90

D. 120
